ACID, ACID EVERYWHERE

A Problem-Based Unit

The College of William and Mary
School of Education
Center for Gifted Education
Williamsburg, Virginia 23185

ACID, ACID EVERYWHERE

A Problem-Based Unit

The College of William and Mary
School of Education
Center for Gifted Education
Williamsburg, Virginia 23185

Center for Gifted Education Staff:
Project Director: Dr. Joyce VanTassel-Baska
Project Managers: Dr. Shelagh A. Gallagher
Dr. Victoria B. Damiani
Project Consultants: Dr. Beverly T. Sher
Linda Neal Boyce
Dana T. Johnson
Donna L. Poland

Teacher Developer:
Sandra Coleman

funded by Jacob K. Javits,
United States Department of Education

KENDALL/HUNT PUBLISHING COMPANY
4050 Westmark Drive   Dubuque, Iowa 52002

Copyright © Center for Gifted Education

ISBN 0-7872-2468-5

Kendall/Hunt Publishing Company has the exclusive rights to reproduce this work,
to prepare derivative works from this work, to publicly distribute this work,
to publicly perform this work and to publicly display this work.

All rights reserved. No part of this publication may be reproduced,
stored in a retrieval system, or transmitted, in any form or by any
means, electronic, mechanical, photocopying, recording, or otherwise,
without the prior written permission of Kendall/Hunt Publishing Company.

Printed in the United States of America
10 9 8 7 6 5 4

CONTENTS

PART I: INTRODUCTORY FRAMEWORK

Introduction   3

Rationale and Purpose   3

Goals and Outcomes   3

Assessment   5

Safety Precautions to be Taken in the Lab   6

Materials List   7

Lesson Flow Chart   7

Tailoring *Acid, Acid, Everywhere* to Your Location   7

Glossary of Terms   9

Letter to Parents   11

PART II: LESSON PLANS

Lesson 1: Introduction to the Problem   15

Lesson 2: Re-Routing Traffic   27

Lesson 3: Hazards and the Map   33

Lesson 4: Introducing the Systems Concept   41

Lesson 5: Playing with pH   49

Lesson 6: Neutralizing an Acid   57

Lesson 7: Diluting an Acid   73

Lesson 8: Introducing the Creek Ecosystem   87

Lesson 9: Measurement of Water Flow   101

Lesson 10: The Effect of Acid on Plants   115

Lesson 11: The Effect of Acid on Materials   129

Lesson 12: Consultation with a Hazardous Waste Expert   131

Lesson 13: Transport Regulations   137

Lesson 14: Resolution and Discussion   141

Lesson 15: Final Overall Unit Assessment Activity   149

Scoring Protocol   157

PART III: REFERENCES   165

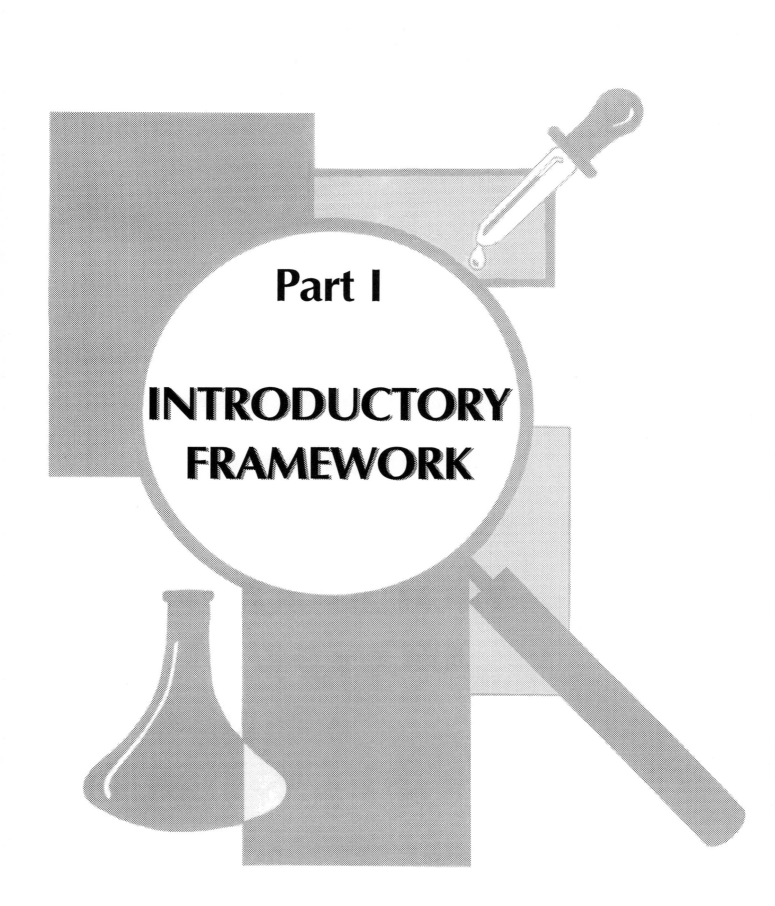

INTRODUCTION

Acid, Acid Everywhere is a problem-based science unit designed for high ability learners which has been successfully used with all learners in a wide variety of situations, from pull-out programs for gifted learners to traditional heterogeneously grouped classrooms. It allows elementary students to explore acid/base chemistry in a novel way, namely through the process of grappling with an ill-structured, "real-world" problem.

Because the unit is problem-based, the way in which a teacher implements the unit will necessarily differ from the way in which most traditional science units are used. Preparing for and implementing problem-based learning takes time, flexibility, and a willingness to experiment with a new way of teaching.

The total time required for completion of *Acid, Acid Everywhere* should be minimally 30 hours, with more time required for additional activities.

RATIONALE AND PURPOSE

This unit has been designed to introduce fourth through sixth grade students to acid/base chemistry in an engaging fashion. The problem-based learning format was chosen in order to allow students to acquire significant science content knowledge in the course of solving an interdisciplinary, "real-world" problem. This format requires students to analyze the problem situation, to determine what information they need in order to come up with solutions, and then to find that information in a variety of ways. In addition to library work and other information-gathering methods, students, with teacher facilitation, will perform experiments of their own design in order to find information necessary to come up with and evaluate solutions to the problem. The problem-based method also allows students to model the scientific process, from the problem-finding and information-gathering steps to the evaluation of experimental data and the recasting or solution of the problem. Finally, the overarching scientific concept of systems provides students with a framework for the analysis of both their experiments and the problem as a whole.

GOALS AND OUTCOMES

➡ To understand the concept of systems

Students will be able to analyze several systems during the course of the unit. These include the "problem system," defined by the boundaries of the area affected by the acid spill; the stream ecosystem (into which the acid flows); and the transportation system (which is disrupted by the acid spill.) In addition, all experiments set up during the course will be treated as systems.

Systems Outcomes

A. For each system, students will be able to use appropriate systems language to identify boundaries, important elements, input, and output.

B. Students will be able to analyze the interactions of various system components with each other and with input into the system, both for the real-world systems and for the experimental systems.

C. Based on their understanding of each system's functioning, students will be able to predict the impact of various approaches to the clean up of the acid spill on each relevant system.

D. Students will be able to transfer their knowledge about systems in general to a newly encountered system. In the final assessment activity, students will be given a new system to analyze in the same way that they have analyzed the systems in the unit.

> ➡ To design scientific experiments necessary to solve given problems

In order to solve these scientific problems, students will be able to design, perform, and report on the results of a number of experiments.

Scientific Process Outcomes

A. Students will be able to explore a new scientific area, namely acid/base chemistry.

B. Students will be able to identify meaningful scientific problems for investigation during the course of working through the acid-spill problem and its ramifications. These problems include the effects of acids on nonliving and living materials, the properties of acids, and so on.

C. During their experimental work, students will:

—Demonstrate good data-handling skills

—Analyze any experimental data as appropriate

—Evaluate their results in light of the original problem

—Use their enhanced understanding of the area under study to make predictions about similar problems whose answers are not yet known to the student

—Communicate their enhanced understanding of the scientific area to others

> ➜ To use the principles of acid/base chemistry

Students will be able to predict the effects of acids and bases on various living and nonliving materials.

SPECIFIC CONTENT OUTCOMES

A. Students will be able to draw the pH scale and correctly indicate where the acidic end, the basic end, and the neutral pH lie on the scale.

B. Students will be able to list some common, household-type acids and bases.

C. Students will be able to devise a safe method for the determination of the pH of an unknown solution.

D. Students will be able to devise a safe method that would allow them to neutralize an acid.

E. Students will be able to devise a safe method that would allow them to determine how much of a known base would be necessary to neutralize a known amount of an unknown acid; in other words, they should be able to construct and use a titration curve.

F. Students should be able to describe what happens to the pH of an acid as it is progressively diluted with water.

ASSESSMENT

This unit contains many assessment opportunities that can be used to monitor student progress and assess student learning. Opportunities for formative assessment include:

- The student's problem log, a written compilation of the student's thoughts about the problem. Each lesson contains suggested questions for students to answer in their problem logs. The problem log should also be used by the student to record data and new information that they have obtained during the course of the unit.

- Experimental design worksheets, which can be used to assess a student's understanding of experimental design and the scientific process, as well as to record information about what was done and what was found during student-directed experimentation.

- Other forms, such as the Traffic Detour Report Form, which are used to help the student explain their solutions to particular parts of the problem.

- Teacher observation of student participation in large-group and small-group activities.

Opportunities for cumulative assessment include:

- The final resolution activity, which involves a small group presentation of a solution to the unit's ill-structured problem; the quality of the solution will reflect the group's understanding of the science involved as well as the societal and ethical considerations needed to form an acceptable solution. This activity requires each group to prepare a written report detailing and justifying their solution, which can be used for assessment purposes as well.
- Final unit assessments, which allow the teacher to determine whether individual students have met the science process, science content, and systems objectives listed in the Goals and Objectives section at the beginning of the unit.

SAFETY PRECAUTIONS TO BE TAKEN IN THE LAB

As this unit involves laboratory work, some general safety procedures should be observed at all times. Some districts will have prescribed laboratory safety rules; for those that do not, some basic rules to follow for this unit and any other curriculum involving scientific experimentation are:

1. Students must behave appropriately in the lab. No running or horseplay should be allowed; materials should be used for only the intended purposes.
2. No eating, drinking, or smoking in lab; no tasting of laboratory materials. No pipetting by mouth.
3. If students are using heat sources, such as alcohol burners, long hair must be tied back and loose clothing should be covered by a lab coat.
4. Fire extinguishers should be available; students should know where they are and how to use them.

Some specific safety rules relevant to implementing this unit:

1. This unit covers acid/base chemistry. Students should not be allowed to use concentrated HCl or to use dangerous household acids and bases such as bleach or oven cleaner in this lab. Any household item that carries a warning label should be considered off limits for student use (although they should be encouraged to look at the warning labels on such products in their study of acids and bases). Food-type acids and bases such as vinegar, lemon juice, and baking soda, are safe for student use; moreover, use of such acids and bases does not require gloves or a lab apron.
2. Safety goggles should be worn at all times in the lab; some provision for washing chemicals out of eyes should be made (either a sink with an eyewash attachment or a large squirt bottle full of water should be available).

3. When diluting a concentrated acid or base: add the concentrated acid or base slowly to water to minimize the likelihood of splashing the acid or base.

4. Do not mix concentrated acids with concentrated bases, as the resulting reaction may be vigorous and dangerous. Do not mix dangerous household chemicals.

MATERIALS LIST

Materials needed for each individual lesson are listed in the "Materials and Handouts" section of the lesson.

LESSON FLOW CHART

Problem-based learning is not easy to plan, because it is driven by student questioning and interest. We have included estimated durations for each lesson in this unit, but be prepared to be flexible and to move with the students. We have also included a diagram (Figure 1) which shows the relationship between the individual lessons and experiments suggested in the unit. In general, lessons shown higher in the diagram are prerequisites for those shown lower in the diagram. Be aware that this diagram may not reflect all of the time that you will need to spend; students may well come up with unanticipated, yet valid, experiments or lines of questioning.

We feel that some of the lessons are essential for all students, while others can be done with a subgroup as long as the subgroup reports its results back to the whole class.

TAILORING *ACID, ACID EVERYWHERE* TO YOUR LOCATION

Classroom experience during the unit piloting process has shown that this unit is *much* more powerful when tailored for the location in which it is being presented. Accordingly, this new version of the unit has been written to help you:

1. Modify the problem statement so that the acid spill occurs at a real site in your area (any highway bridge over a river or stream will do).

2. Use local highway, topographic, and resource maps in place of the sample artificial maps supplied in the unit. This customization has several potential advantages: the site is familiar to the students, there are real, detailed topographic and street maps available, and the students can talk to the actual people who would be involved in managing such a spill.

3. Involve local experts (hazmat teams, emergency planners, etc.) as speakers and on-going resources in the problem-solving process.

4. Visit the site of the purported spill; observe the local ecosystem and sample the water.

FIGURE 1

LESSON FLOW CHART

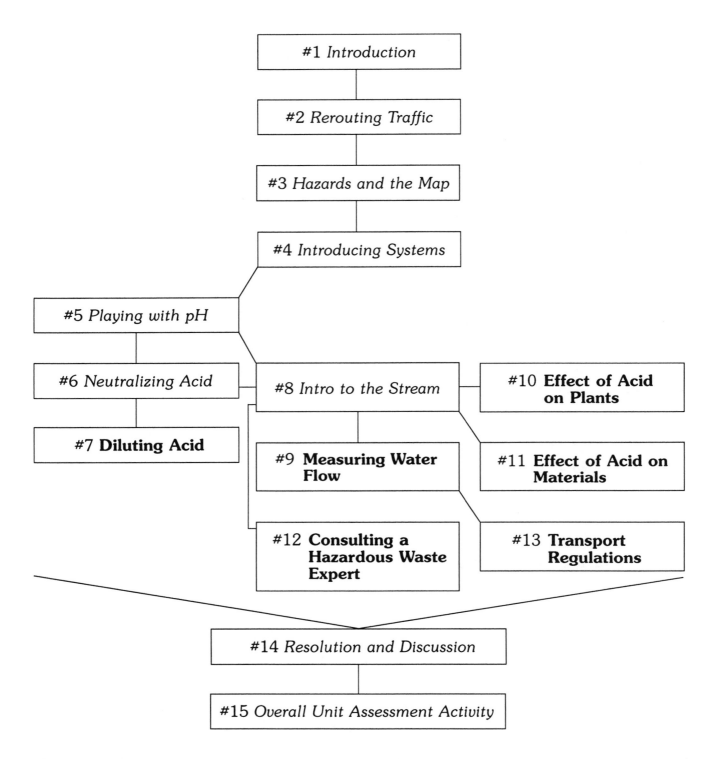

ITALICS: Represents lessons that are essential for all students.
BOLD: Represents lessons that can be done by a subgroup and reported out.

5. Work with librarians to plan the unit and to assist students in finding information. In addition to school librarians and academic librarians, special libraries (museums, corporations, historical societies, etc.) offer vast resources relevant to the unit.

GLOSSARY OF TERMS

Abiotic: Nonliving component of the environment including soil, water, air, light, nutrients, and the like.

Acidic Solution: A solution containing more hydronium than hydroxide ions. On a pH scale, these solutions would have a pH value from 0–6.

Base Solution: A solution containing more hydroxide ions than hydronium ions. On the pH scale, these solutions would have a pH value from 8–14.

Biotic: The living component of the environment.

Boundary (Systems): Something that indicates or fixes a limit on the extent of the system.

Dilution Process: The process through which a solution is diminished in strength, flavor, or brilliance by the addition of another solution.

Element (Systems): A distinct part of the system; a component of a complex entity (system).

Input (Systems): Something that is put in the system; an addition to the components of the system.

Litmus Paper: Paper (unsized) colored with litmus (a dye) and used as an indicator for acids/bases; turns red in acid solutions and blue in alkaline solutions.

Neutralization Process: The process through which something made chemically neutral; an acid or a base is brought to a neutral pH of 7.

Output (Systems): Something that is produced by the system; a product of the system interactions.

pH Indicators: Strips of specially treated paper that change colors when dipped in a solution, indicating a pH reading between 0 and 14. These give a more precise reading than litmus paper.

pH Neutral: Represented by a pH value of 7 on the pH scale, a solution that is pH neutral has an equal number of hydroxide and hydronium ions.

Scientific Process (or Research): The scientific research process can be described by the following steps:

1. Learn a great deal about your field.
2. Think of a good (interesting, important, and tractable) problem.
3. Decide which experiments/observations/calculations would contribute to a solution of the problem.
4. Perform the experiments/observations/calculations.
5. Decide whether the results really do contribute to a better understanding of the problem. If they do not, return to either step 2 (if you're very discouraged) or step 3. If they do, go to step 6.
6. Communicate your results to as many people as possible. If they're patentable, tell your lawyer before you tell anyone else, and write a patent application or two. Publish them in a scientific journal (or if they are really neat, in *The New York Times*); go to conferences and talk about them; tell all of your friends.

System: A group of interacting, interrelated, or interdependent elements forming a complex whole.

Titration: A method of reacting a solution of unknown concentration with one of known concentration; this procedure is often used to determine the concentrations of acids and bases.

LETTER TO PARENTS

Dear Parent or Guardian:

Your child is about to begin a science unit that uses an instructional strategy called problem-based learning. In this unit students will take a very active role in identifying and resolving a "real world" problem constructed to promote science learning. Your child will not be working out of a textbook during this unit but will be gathering information from a variety of other sources both in and out of school.

The goals for the unit are:

- *To understand the concept of "systems."*

    Students will be able to analyze several systems during the course of the unit. These include the "problem system," defined by the boundaries of the area affected by the acid spill; the stream ecosystem; and the transportation system. In addition, all experiments set up during the course will be treated as systems.

- *To use the principles of acid/base chemistry.*

    Students will be able to predict the effects of acids and bases on various living and nonliving materials.

- *To design scientific experiments necessary to solve given problems.*

    In order to solve scientific problems, students need to be able to design, perform, and report on the results of a number of experiments. During their experimental work, students will:

    —Demonstrate good data-handling skills

    —Analyze any experimental data as appropriate

    —Evaluate their results in the light of the original problem

    —Use their enhanced understanding of the area under study to make predictions about similar problems whose answers are not yet known to the student

    —Communicate their enhanced understanding of the scientific area to others

Since we know from educational research that parental involvement is a strong factor in promoting positive attitudes toward science, we encourage you to extend your child's school learning through activities in the home.

Ways that you may wish to help your child during the learning of this unit include:

- Discuss systems, including family systems, educational systems, etc.
- Discuss the problem they have been given.

- Engage your child in scientific-experimentation exercises based on everyday events such as: In a grocery store, how would you test whether it's better to go in a long line with people having few items or a short line with people having full carts?
- Take your child to area science museums and the library to explore how scientists solve problems.
- Use the problem-based learning model to question students about a question they have about the real world, e.g., How does hail form? Answer: What do you know about hail? What do you need to know to answer the question? How do you find out?

Thank you in advance for your interest in your child's curriculum. Please do not hesitate to contact me for further information as the unit progresses.

Sincerely,

Part II

LESSON PLANS

Lesson 1: Introduction to the Problem

LESSON LENGTH Two sessions

INSTRUCTIONAL PURPOSE

- To introduce students to problem statement to be explored throughout the unit.
- To address scientific process goals by beginning the use of problem-based learning.

MATERIALS AND HANDOUTS

Handout 1.1: Problem Statement (to be localized by the teacher)
Handout 1.2: "Need to Know" board on chalkboard or butcher paper
Handout 1.3: Problem Log Questions
Handout 1.4: Smudged Invoice

Session 1

THINGS TO DO

1. Read the problem statement to students (Handout 1.1); provide them with their own copies as well.

2. Help students sort out information and questions derived from the problem statement into three categories on the "Need to Know" board (Handout 1.2): What We Know, What Do We Need To Know, and How Can We Find Out.

3. Have students identify key words and phrases as they organize the elements of the problem.

4. As students generate questions for the "Need to Know" board (Handout 1.2), ask them to tell why the information is important or what idea they are pursuing by asking for the information.

5. Help students prioritize the "Need to Know" board (Handout 1.2) from most to least critical. Debate reasons for prioritizing choices.

6. Ask students to identify resources that will help them answer or further investigate the elements of the "Need to Know" board (Handout 1.2). Divide the learning issues among students so that each student or small group of students will bring information to class for the next session.

THINGS TO ASK

PROBLEM-RELATED QUESTIONS

- What's going on at the bridge?
- What are we, as State Highway Patrol supervisors, supposed to do about the situation?
- What seem to be the key pieces of information?
- What seems to be the main problem?
- Are there other problems?
- Where can we find answers to these questions?
- Do you have any ideas right now about what to do about the situation?

HANDOUT 1.1

SAMPLE INITIAL PROBLEM STATEMENT

You are the supervisor of the day shift of the State Highway Patrol. It is 6:00 AM on a cool autumn morning. You are sleeping when the phone rings. You answer and hear, "Come to the (Clear Creek bridge on Route 15)*. There has been a major accident and you are needed."

Quickly you dress and get on the road to hurry to the site of the emergency. As you approach the bridge, you see an overturned truck that has apparently crashed through the metal guard rail. It has lost one wheel and is perched on its front axle. You see "corrosive" written on a small sign on the rear of the truck. There is a huge gash in the side of the truck and from the gash a liquid is running down the side of the truck, onto the road, and down the hill into a creek. Steam is rising from the creek. All traffic has been stopped and everyone has been told to remain in their cars. Many of the motorists trapped in the traffic jam appear to be angry and frustrated. Police officers, firemen, and rescue squad workers are at the scene. They are all wearing coveralls and masks. The rescue squad is putting the unconscious driver of the truck onto a stretcher. Everyone seems hurried and anxious.

We recommend that this statement be localized to your area. See Introduction and Teacher's Guide for specific details.

HANDOUT 1.2*

NEED TO KNOW BOARD

What do we know?	What do we need to know?	How can we find out?

*The student Need to Know boards should be posted around the room and used as a working document throughout the unit as students answer some questions and create new ones.

Session 2

THINGS TO DO

1. Have students report on the information they found overnight. Use this new information to modify the "Need to Know" board as necessary, shifting questions from the "What Do We Need to Know" category to the "What Do We Know" category as appropriate.

2. Provide the smudged invoice for students to analyze. Have them identify key new pieces of information on the invoice and modify the "Need to Know" board accordingly.

3. Ask students what they are going to need to know in order to solve the problem and write their answers on the "Need to Know" board.

4. Prioritize the list based on negotiations with students; assign information-gathering tasks.

THINGS TO ASK

- What questions are answered by the new information?
- What questions do we still have to answer?
- What new questions do you have?
- What are the effects of acid?
- Will acid affect only the things it touches?
- What are the things about which we may have to learn to solve the problem?
- Is the problem different today than it was yesterday?
- How are we going to solve this problem?
- What sort of strategies should we use?
- What will the different stages of our job be?
- How will you know when the problem is solved?

ASSESSMENT

Have students answer the questions posed in Handout 1.3 in their Problem Logs.

NOTE TO TEACHER

1. The people responsible for coordinating the emergency response and cleanup efforts will vary from location to location. For maximum verisimilitude, find out whose responsibility this would be in your area and modify the job title in the problem statement accordingly.

2. While the spill is concentrated HCl, it is advisable not to let the students know this until after Lesson 7 in the unit.

Handout 1.3
Problem Log Questions

At the end of Session 2, have students paraphrase the problem situation in the Problem Log and answer the following questions:

1. After our first two days of discussion, what do you think the problem really is?

2. Why do you think this is the main problem?

3. Is it the same problem you thought it was when we first started talking?

4. How has it changed?

5. What are the issues you are most interested in finding out about?

Handout 1.4
Smudged Invoice

Invoice #3944

Coleman of Virginia

Account # 123456	Sold By CMW	Reference # 2284	Ship Via UPS	Terms Net 30	Date 12-1-95
Qty. Ordered	Qty. Shipped	Item #	Description	Unit Price	Total Price
				Sale Amount	
				Sales Tax	
				Freight	
				Total	

Please enter our order for the above merchandise subject to the terms and conditions on the reverse side of this order.

Lesson 2: Re-Routing Traffic

LESSON LENGTH: One–two sessions

INSTRUCTIONAL PURPOSE

- To generate additional questions about the spill through continuing the problem-based learning process.

MATERIALS AND HANDOUTS

Local maps of accident site (topo maps, highway maps, ortho-photo maps)

Handout 2.1: "Need to Know" board

Handout 2.2: Traffic Detour Report Form

New information: Communication from highway patrolman

THINGS TO DO

1. Present new information. New information: *Can you meet in fifteen minutes? Sir/Madam—We need your input on a safe detour.* Modify "Need to Know" board.

2. When prompted by student request, present students with a highway map of the accident site.

3. Update the "Need to Know" board (Handout 2.1) to accommodate the new information and questions generated by the map.

4. Divide the class into small groups. Ask each group to develop an acceptable detour route to allow traffic to safely proceed around the site of the spill for the duration of the clean-up period. Supply them with Traffic Detour Report forms (Handout 2.2) saying "Sargeant _____ (name) will have to write up the detour on one of these. To help him/her out, let's fill out these forms with our recommendations."

5. After giving students time to complete their forms, have the groups discuss their planned detour routes and come up with a class consensus route.

6. Groups should next be told that another of their responsibilities as Highway Patrol Supervisors is to complete the "Hazard Alert" Form (Handout 3.3). Give each group a form and have them fill it out with information they have gleaned relative to the accident.

THINGS TO ASK

- What are the things you need to pay attention to when you re-route the traffic?
- Who is going to be affected most by the re-routing?
- Does any special group need to be accommodated? Does any group deserve special treatment? Why? Why not?
- Should everyone be re-routed the same way?
- How will you go about planning your alternate route?
- What else do you notice in the map?
- What are the other significant pieces of information on the map?
- What do you notice about their relationship to the accident?

NOTE TO TEACHER

If you have tailored the unit to your area, use appropriate local maps for this lesson.

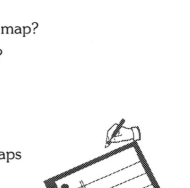

HANDOUT 2.1*

NEED TO KNOW BOARD

What do we know?	What do we need to know?	How can we find out?

*The student Need to Know boards should be posted around the room and used as a working document throughout the unit as students answer some questions and create new ones.

Handout 2.2

Traffic Detour Report Form

Name of Highway Patrol Officer(s): _____

Date of Report: _____

Please consult with Highway Patrol Officers and make a detour route recommendation for the flow of traffic around the accident. Write up the exact route of your detour, then complete the rest of this report. Be prepared to defend your decision.

1. What are the positive aspects of your detour?

2. What are the negative aspects of your detour?

3. How will citizens respond to your choice? Who will support it? Who won't?

4. Why should traffic be detoured this way?

lesson 3
Hazards and the Map

LESSON LENGTH: One session

INSTRUCTIONAL PURPOSE
- To help students understand that acid poses specific hazards for specific groups.

MATERIALS AND HANDOUTS

Local maps, information about land use in spill area, locations of schools, industry, housing, etc.

Handout 3.1: "Need to Know" board
Handout 3.2: Weather Report
Handout 3.3: Hazard Alert Forms

THINGS TO DO

1. Distribute topographic maps and weather information. Have student groups determine direction of water flow in the creek and wind flow around the site of the spill and record their predictions in their Problem Logs. Each group should be prepared to report how they determined current direction and wind direction.

2. Reconvene groups to see what additional hazards have been posed by the new information about current and wind direction. What does this mean for the problem? Use this information to update the "Need to Know" board.

3. Tell student groups that another of their responsibilities as Highway Patrol Supervisors is to identify and inform individuals or groups who should be contacted regarding the accident.

4. Using the maps, have students make a class list of the individuals or groups of people who might be affected by the spill.

5. After students identify the different groups which would need to be contacted regarding the accident, divide the class into groups who are responsible for dealing with the affected individuals or organizations. Students should complete Hazard Alert Forms to be communicated to their special interest group; details of the specific hazard posed by the spill to each interest group should be listed on the forms.

THINGS TO ASK

- What do you notice about the spill from the information on these maps?
- What are some questions the maps raise about the spill?
- Does it matter which side of the bridge the spill was on? Why?
- If the acid spills in the water, will it affect the entire creek? Why?
- Which areas adjacent to the creek will be affected? How do you know?
- What could some other possibilities be?
- How could we find out for sure?
- What about the fumes coming from the water? What will they do to the surrounding area?
- Could normal weather patterns create problems here? How?
- What might the Highway Patrol have to do?
- Should action be taken to inform anyone about the problem? Who?
- How would this information help us to decide who to inform about the spill? How could we find out?

ASSESSMENT

Use of Problem Log entries should be judged according to the following criteria:

1. Ability to determine current and wind direction.
2. Quality of issues raised by interest groups as related specifically to current and wind direction.
3. Quality of group reports.
4. Accuracy and thoroughness of hazard reports.

Handout 3.1

Need to Know Board

What do we know?	What do we need to know?	How can we find out?

HANDOUT 3.2

WEATHER REPORT

After the early morning fog burns off, today will be clear, with winds from the west at 5–10 mph. The high will be about 85 degrees, with the overnight low between 55 and 60 degrees. Tomorrow's high will be 85 degrees, with winds gusting from the west at 5 to 10 mph.

HANDOUT 3.3
HAZARD ALERT FORM

Highway Patrol Officer: _____

Date of report: _____

Specific hazard posted: _____

Site(s) which need to be posted: _____
_____
_____
_____

Reasons for posting site: _____
_____
_____
_____

Who needs to be informed? Why? _____
_____
_____
_____

What are the possible effects of the hazard? _____
_____
_____
_____

lesson 4

Introducing the Systems Concept

LESSON LENGTH: One session

INSTRUCTIONAL PURPOSE
- To introduce the systems concept and allow students to apply it to the problem situation.

MATERIALS AND HANDOUTS
Handout 4.1: State Diagram
Handout 4.2: Systems Parts Chart
Handout 4.3: System Diagram

THINGS TO DO

1. Tell the students that you're taking a time out from the problem.

2. Describe the concept of "systems" to students. Introduce students to the terms associated with systems and have students derive or find definitions for each term:

   - boundary: determines what is inside the system and what is outside the system
   - elements: the parts that make up the system
   - input: anything that goes into the system from outside the system
   - output: anything that the system releases to the outside world
   - interactions: the effects that parts of the system have on each other

3. Hand out the State Diagram (Handout 4.1). Ask students to diagram the system and fill out a parts chart. Debrief.

4. Ask them if the problem situation could be considered as a system. Give the students copies of the highway map; ask them to draw the system's boundaries. As a group, discuss the boundaries, elements, input, and output of the problem spill system.

THINGS TO ASK

For the acid spill system (for each system analyzed):

- What would they define as its boundaries? Its elements?
- Why did they pick these boundaries? Elements?
- Would they want to define the acid spill itself as an element or as input? Why or why not?
- What are the possible interactions of the acid with the other system elements?
- Does the system produce any output?
- Does everyone agree that these have to be the boundaries, or are the boundaries arbitrary?
- What kind of output does it produce, and what interactions supply this output?
- How might considering the spill situation as a system help us to solve the problem?
- What other systems do you see affected by the acid spill?
- How would knowing about these other systems help us to assess the impact of the acid spill?

ASSESSMENT

1. State diagrams and parts charts.
2. Have students fill out a system diagram and systems parts chart for the spill system in their Problem Logs.

HANDOUT 4.1
STATE DIAGRAM

The outline of the State of Maryland may be found on the next page. Identify some of the major elements, inputs, outputs, boundaries, and interaction effects for this system.

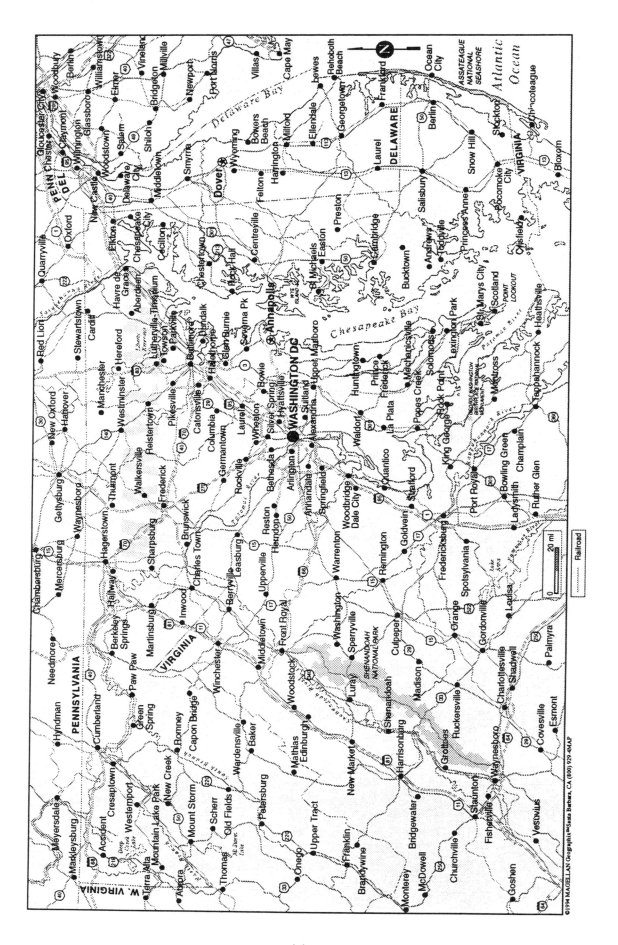

HANDOUT 4.2

Problem Log: System Parts Chart

1. What are the boundaries of the system? Why did you choose them? Were there other possibilities?

2. List some important elements of the system.

3. Describe input into the system. Where does it come from?

4. Describe output from the system. What part(s) of the system produce it?

5. Describe some important interactions:
    a. among system elements

    b. between system elements and input into the system

6. What would happen to the system if the interactions in 5a could not take place? In 5b?

HANDOUT 4.3

Problem Log: System Diagram

Draw a picture of the system. Show where its boundaries are. Label and describe all of its parts.

Lesson 5: Playing with pH

LESSON LENGTH: One–two sessions

INSTRUCTIONAL PURPOSE

- To introduce and apply the pH scale in experimentation.

MATERIALS AND HANDOUTS

Cups (3–4 small cups for each liquid used)

Safety goggles, aprons, paper towels

A collection of safe kitchen-type liquids, such as:

Vinegar	Antacid tablet (dissolved in water)
Apple juice	Shampoo
Orange juice	Pepsi/Coke
Steeped tea leaves	Tap water
Cream of tartar	Distilled or deionized water
Liquid dish detergent	
Baking soda (dissolved in water)	
Aspirin (dissolved in water)	

pH Fix universal indicator sticks (available from Edmund Scientific, 1-609-547-8880; catalog # Y36, 302: do NOT use litmus paper)

Handout 5.1: pH Scale and pH Data Sheet

Handout 5.2: Problem Log

THINGS TO DO

1. Pass out pH sticks and ask students what they notice. Ask students what they think the pH stick does. Ask them what they think the scale on the side of the box containing the sticks means. Tell students the pH sticks were used at the accident site to test the liquid in the spill.

2. Pass out a copy of the Handout 5.1 (the pH scale and Data Table). Point out the correspondence between the scale on the side of the pH stick box and the scale on the handout.

3. Talk to students about safety procedures for the lab activity. Show students the substances that they will be working with during the lab activity. Ask students to list precautions that would protect their skin, eyes, and clothing; these precautions should include both proper behavior during the activity and necessary safety equipment. Write the list of "lab safety rules" on the chalkboard or butcher paper so everyone can see it during the lab.

4. Put a small amount of each liquid in a cup for each testing station (there should probably be three or four stations, depending on the number of students you want working together). Mark all cups with the name of the substance.

5. Give each group of students a box of pH sticks and copies of the Handout 5.1. Ask students to determine the pH of each substance and record it on their data sheet. Allow students to discuss their results with each other. **Teachers should monitor this activity closely to ensure that students combine only appropriate amounts of substances.**

6. When students have finished their basic testing, give them a challenge: Can they come up with a way of neutralizing the lemon juice? (They should be able to do this by adding baking soda.) (Ask them to first write down their predictions about the challenges and then test them to see if the predicted methods work.)

7. After the activity is over, bring students back together and discuss the results. Hand out the questions to be answered in the Problem Log (Handout 5.2).

THINGS TO ASK

1. Before the activity:
   - What is this? What could it be used for?
   - What do you see on the side of the sheet?
   - What do you think the scale is used for?

2. After the activity:
   - What substance was the most acidic?
   - Was this substance as acidic as the run-off from the spill?
   - What substance was the most basic?

- Were there any neutral substances?
- Did any of your results surprise you?
- Could anyone manage to come up with a method to meet the first challenge? The second? How did you do it?
- Can you define now the terms acid, base, and neutral? (Refer students to unit glossary as needed.)

3. Problem-related questions:
   - How could this information help us figure out the problem?
   - What would you think would happen to living things exposed to the runoff from the spill? How could you find out?
   - Is the runoff from the spill safe?
   - What else might be affected by the runoff from the spill?
   - What new questions should we add to the "Need to Know" board?

ASSESSMENT

Problem Log entries.

EXTENSIONS

Have students go home and read labels on food items and on cleaning items (such as oven cleaner). Caution students about safety. Have them look for warning information, such as "contains a strong acid"; precautions for use; and other safety-related information. Also, have them make a list of all of the ingredients that they can identify as acids or bases (citric acid, for example). Students could also look up the chemical properties of some of these substances in The Merck Index, which lists thousands of chemicals and their properties.

NOTE TO TEACHER

For this lesson and other experiments in the unit, a useful baking soda stock solution can be made by mixing (forever) a quarter cup of baking soda with two and a half cups of warm water (or just enough to get all of the baking soda to dissolve). You may want to warm it in a microwave oven to speed up the process. If the students prepare the solution, be sure that they notice that the water gets colder as the baking soda dissolves: this contrasts nicely with the warming that occurs with the mixing of the baking soda solution and vinegar. It is important for them to notice the temperature changes that occur, because they can affect the viability of different solutions to the problem. Dropping large amounts of baking soda solution on the acid spill will neutralize the acid but also generate a lot of heat in the process.

It would be wise at this time to begin the experiment for Lesson 10 as it will take three weeks to see results.

Handout 5.1
pH Scale

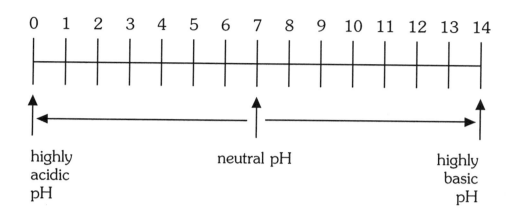

pH Scale Data Sheet

DATA SHEET	
Substance	pH

HANDOUT 5.2

PROBLEM LOG QUESTIONS

1. Describe in your own words what the pH scale measures. How is this useful to people studying pollution? How would you want to use it in your investigation of the corrosive spill?

2. List and describe the safety procedures we discussed in class today. Include both safety equipment and safe behaviors. After today's activity, do you think we need to add more to the list? Why or why not? If yes, what are they? Are any of our rules unnecessary? If yes, which and why?

Teacher Background Information on pH

The pH scale is a logarithmic (exponential) scale rather than linear scale. The numbers on the pH scale represent the negative logarithm (exponent) of the concentration of hydrogen ions, which is expressed in moles of hydrogen ions per liter of solution. This means that the concentration can be represented as fractions of a mole as noted below.

pH Scale

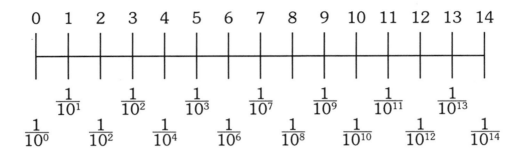

Note that the highest concentrations of hydrogen ions are found in solutions with low numbers on the pH scale. Each move up the scale represents a solution that is ten times less concentrated. Students may think that the presence of lots of hydrogen ions is either good or bad. It is neither! They should understand that pH is just a measure of how concentrated these ions are. Solutions with various concentrations have different properties.

lesson 6: Neutralizing an Acid

LESSON LENGTH: 3 sessions

INSTRUCTIONAL PURPOSE
- To introduce students to experimental design.

MATERIALS AND HANDOUTS

Distilled water (1 gallon)
White vinegar
Baking soda
Clear plastic cups (2 per group)
pH sticks
Calibrated eyedroppers (like those used for infant Tylenol)
Safety equipment: goggles, aprons, gloves
Thermometers (1 per group)
Graph paper

Handout 6.1: Student Brainstorming Worksheet
Handout 6.2: Student Experiment Worksheet
Handout 6.3: Student Protocol Worksheet
Handout 6.4: Laboratory Report Form
Handout 6.5: Problem Log Questions

Session 1: Planning the Experiment

THINGS TO DO

1. Review the results of Lesson 5. Ask students if anything that they have learned could help them with the problem-related question of how to go about cleaning up the acid spill. Remind them of the results of the challenges at the end of the lab session in Lesson 5. They should have come up with the fact that if

you add base to an acid, the resulting solution has a pH closer to neutrality than those of the starting materials.

2. Show them the materials for this lesson. Break students into small groups, and ask them to come up with a method for raising the pH of the vinegar to 7 using the materials available. Tell them that they need to be able to write this method down in such a way that other people could repeat it. Pass out a copy of the Student Brainstorming Worksheet (Handout 6.1) and have them fill it out.

3. Discuss students' brainstorming results as a class.

4. Next, pass out copies of the Student Experiment Worksheet (Handout 6.2). If this is the first time students have seen it, have the class fill it out.

5. After students have filled out the Student Experiment Worksheet (Handout 6.2), have each group write their experimental protocol on the Student Protocol Worksheet (Handout 6.3). If time is running short, this could be assigned as homework. They should include every step that they plan to take in the experiment, the materials they will use, and a data table in which to record their data.

THINGS TO ASK

Before the experiment:

- Can we alter the pH of the vinegar?
- How could we do it?
- Could we control whether or not the pH changed a little or a lot? Or would the change always be the same?
- How would we go about testing these ideas?

NOTE TO TEACHER

Sample Protocol for Neutralizing an Acid

1. Measure out 5 ml of the vinegar into a plastic cup.
2. Begin by adding 1 ml of the baking soda solution to the vinegar. Stir. Record the pH and temperature of the solution in the cup. Repeat this process several times, recording pH and temperature after each addition.
3. When the pH reaches 7, the vinegar has been neutralized. Continue adding baking soda solution until pH reaches 8 or 9.
4. Graph the pH of the solution versus the amount of baking soda added. This graph is called a **titration curve** (see glossary).

Handout 6.1
Student Brainstorming Worksheet

1. What do we need to find out? (What is the scientific problem?)

2. What materials do we have available?

3. How can we use these materials to help us find out?

4. What do we think will happen? (What is our hypothesis?)

5. What will we need to observe or measure in order to find out the answer to our scientific question?

Adapted from Cothron, J. G., Giese, R. N., & Rezba, R. J. (1989). *Students and research.* Dubuque, IA: Kendall/Hunt Publishing Co.

Handout 6.2

Student Experiment Worksheet

Title of Experiment:

Hypothesis (Educated guess about what will happen):

Independent Variable (The variable that **you change**):

Dependent Variable (The variable that responds to changes in the independent variable):

Observations/Measurements to Make:

Constants (All the things or factors that remain the same):

Control (The standard for comparing experimental effects):

Handout 6.3
STUDENT PROTOCOL WORKSHEET

1. List the materials you will need.

2. Write a step-by-step description of what you will do (like a recipe!). List every action you will take during the experiment.

3. What data will you be collecting?

4. Design a data table to collect and analyze your information.

Session 2

THINGS TO DO

1. Check the protocols for safety and workability and have students perform their experiments.
2. Discuss their results.
3. Pass out the Laboratory Report Forms (Handout 6.4). Have them write their lab reports and incorporate them into their Problem Logs.

THINGS TO ASK

1. *After the experiment:*
   - Did your method work?
   - Is there anything you would change about your method if you did it again?
   - Does this information help to clarify the problem? How?
   - I've heard it said that "The solution to pollution is dilution." Does this suggest another method for neutralizing the acid spill?
   - Did you notice anything about the temperature of the vinegar solution as you added base to it? (It should have warmed up.)
   - Does this suggest any possible side effects from cleaning up the acid spill by adding base to it?
   - The acid in the spill is much more concentrated than the vinegar we used in class. What does that suggest about the amount of base we would have to add to the acid in the spill in order to neutralize it?
   - Would it be safe to clean up the acid spill by adding base to it? Why or why not?
   - Do you think that the base you add to the acid has to be dissolved in water, or could you just add the dry chemical itself?
   - Do we need to modify the "Need to Know" board?

2. *Systems-Related Questions:*
   - Was your experimental setup a system?
   - Draw a picture of your setup and label its components.
   - What were its boundaries?

- What were the elements of your experimental setup?
- Did you put anything into your system?
- What was the output from your experimental system?
- How did the elements of your experimental system interact with any input you added? How did the input change your experimental system?
- What interactions inside the system allowed the system to produce output?

ASSESSMENT

All student worksheets may be used to assess individual and group progress.

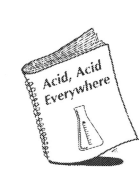

HANDOUT 6.4

LABORATORY REPORT FORM

1. What did you do or test? (Include your experiment title.)

2. How did you do it? What materials and methods did you use?

3. What did you find out? (Include a data summary and the explanation of its meaning.)

4. What did you learn from your experiment?

5. What further questions do you now have?

6. Does the information you learned help with the problem?

Session 3: Debriefing Group Experiments

INSTRUCTIONAL PURPOSE

- To assess student integration of the problem and the concept of systems within their experiments from Session 2.

MATERIALS AND HANDOUTS

Handout 6.5: Student Brainstorming Worksheet

THINGS TO DO

1. Pass out Handout 6.5 and ask students in their groups to respond to the questions posed. (Allow 20 minutes for this activity.)

2. Ask each student group to respond to each question in turn. Discuss student responses and clarify misconceptions as appropriate.

HANDOUT 6.5

GROUP DISCUSSION QUESTIONS AND TASK

1. If you had a sample of the acid from the spill, how could you neutralize it?

2. If you had a sample of the acid from the spill and knew how much acid there was in the whole spill, how could you figure out how much base you would need to neutralize the whole spill?

3. What did you notice about the temperature of the solution of acid as you added more base to it? Would adding base to the spill in order to neutralize it have any side effects?

4. Draw a picture of your experimental setup. Label its boundaries and its elements. List the input you put into your experimental system and the output that came out of it. What interactions inside the system allowed it to produce output? Were there interactions between the original system elements or interactions with input that you added?

Lesson 7

Diluting an Acid

LESSON LENGTH: Two sessions

INSTRUCTIONAL PURPOSE
- To have students investigate what happens to the pH of an acid when it is diluted with water.

MATERIALS AND HANDOUTS

Distilled water (2 gallons)   pH sticks
White vinegar   Graduated cyclinder for each group
Large clear plastic cups   Graph paper
Calibrated eyedroppers (like those used for infant Tylenol)
Safety equipment: goggles, aprons, gloves

Handout 7.1: Student Brainstorming Worksheet
Handout 7.2: Student Experiment Worksheet
Handout 7.3: Student Protocol Worksheet
Handout 7.4: Laboratory Report Form
Handout 7.5: Problem Log Questions

THINGS TO DO

1. Tell students that it's been said that "The solution to pollution is dilution." Ask students what they think would happen if they added a lot of water to an acid.

2. Show them the materials for this lesson. Break students into small groups, and ask them to come up with a method for raising the pH of the vinegar to 7 using the materials available. Tell them that they need to be able to write this method down in such a way that other people could repeat it. Pass out a copy of the Student Brainstorming Worksheet (Handout 7.1) and have them fill it out. When they have finished brainstorming, discuss their ideas.

3. Pass out copies of the Student Experiment Worksheet (Handout 7.2) and have students fill it out. Discuss their finished worksheet with them.

4. Pass out the Student Protocol Worksheet (Handout 7.3) and have each group write their experimental protocol. If time is running short, this could be assigned as homework.

THINGS TO ASK

Before the experiment:

- Can we alter the pH of the vinegar just by adding water?
- Why or why not?
- How could we do it?
- Could we control whether or not the pH changed a little or a lot? Or would the change always be the same?
- How would we go about testing these ideas?

NOTE TO TEACHER

Sample Protocol for Diluting an Acid

1. Measure out 1 ml of the vinegar into a plastic cup.
2. Add 1 ml of amounts of water to the vinegar. Stir. After each addition, record the pH of the solution in the cup.
3. When the pH reaches 7, stop adding water.
4. Graph the pH of the solution versus the amount of water added.

(It will require a lot of water to bring the pH of the solution to neutral. Allow more time for this experiment than the neutralization experiment in Lesson 6.)

Handout 7.1

Student Brainstorming Worksheet

1. What do we need to find out? (What is the scientific problem?)

2. What materials do we have available?

3. How can we use these materials to help us find out?

4. What do we think will happen? (What is our hypothesis?)

5. What will we need to observe or measure in order to find out the answer to our scientific question?

Adapted from Cothron, J. G., Giese, R. N., & Rezba, R. J. (1989). *Students and research.* Dubuque, IA: Kendall/Hunt Publishing Co.

HANDOUT 7.2
STUDENT EXPERIMENT WORKSHEET

Title of Experiment:

Hypothesis (Educated guess about what will happen):

Independent Variable (The variable that **you change**):

Dependent Variable (The variable that responds to changes in the independent variable):

Observations/Measurements to Make:

Constants (All the things or factors that remain the same):

Control (The standard for comparing experimental effects):

Handout 7.3

Student Protocol Worksheet

1. List the materials you will need.

2. Write a step-by-step description of what you will do (like a recipe!). List every action you will take during the experiment.

3. What data will you be collecting?

4. Design a data table to collect and analyze your information.

Session 2

THINGS TO DO

1. Check the protocols and have students perform their experiments. Discuss their results.

2. Have students fill out a Laboratory Report (Handout 7.4) and incorporate it into their Problem Logs.

3. If this experiment was performed by a small group rather than by the whole class, have the group members report their findings to the class as a whole. Have the small group provide copies of all of their worksheets and their data to the other students. Have all students answer the questions in Handout 7.5 in their Problem Logs.

THINGS TO ASK

1. *After the experiment*:

   - Did your method work?
   - Is there anything you would change about your method if you did it again?
   - Does this information help to clarify the problem? How?
   - Did you notice anything about the temperature of the vinegar solution as you added water to it?
   - Does this suggest any possible side effects from cleaning up the acid spill by diluting it?
   - How much water did it take to raise the pH to 7?
   - What does this suggest about using water to clean up the acid spill?
   - Would it be safe to clean up the acid spill by adding water to it? Why or why not?
   - Do we need to modify the "Need to Know" board?

2. *Systems-Related Questions*:

   - Was your experimental setup a system?
   - Draw a picture of your setup and label its components.
   - What were its boundaries?
   - What were the elements of your experimental setup?

- Did you put anything into your system?
- What was the output from your experimental system?
- How did the elements of your experimental system interact with any input you added? How did the input change your experimental system?
- What interactions inside the system allowed the system to produce output?

ASSESSMENT

All worksheets in these two sessions may be used to assess students individually and collectively.

HANDOUT 7.4

LABORATORY REPORT FORM

1. What did you do or test? (Include your experiment title.)

2. How did you do it? What materials and methods did you use?

3. What did you find out? (Include a data summary and the explanation of its meaning.)

4. What did you learn from your experiment?

5. What further questions do you now have?

6. Does the information you learned help with the problem?

HANDOUT 7.5
PROBLEM LOG QUESTIONS

1. If you had a sample of the acid from the spill, how could you dilute it?

2. If you had a sample of the acid from the spill and knew how much acid there was in the whole spill, how could you figure out how much water you would need to dilute all of the acid in the spill?

3. Draw a picture of the experimental setup. Label its boundaries and its elements. List the input you put into your experimental system and the output that came out of it. What interactions inside the system allowed it to produce output? Were there interactions between the original system elements or interactions with input that you added?

Lesson 8: Introducing the Creek Ecosystem

LESSON LENGTH: Two sessions

INSTRUCTIONAL PURPOSE
- To assist students in their analysis of the effect of the acid on the organisms in the creek.

MATERIALS AND HANDOUTS

Field guides to life forms commonly found in streams in your area

Handout 8.1: System Parts Chart
Handout 8.2: System Diagram
Handout 8.3: Communique from (insert your water source's name)
Handout 8.4: Problem Log Questions

Session 1

THINGS TO DO

1. Hand students the "Communique from the Friends of (insert your water source's name)" (Handout 8.3) Environmental Action Group. After they have read it, ask them to modify the "Need to Know" board.

2. Introduce students to the elements of an ecosystem. These include biotic elements (all of the living things present in the ecosystem) and abiotic factors (all of the nonliving characteristics of the ecosystem-water pH, soil characteristics, topography, and so on.)

3. Have students do some brainstorming in small groups about the creek ecosystem. Pass out copies of the System Parts Chart (Handout 8.1) to assist in this process; ask them to fill it in as best they can.

4. Once students have filled in their system parts charts, ask them to fill out a "Need to Know" board form to describe what they don't know about the creek ecosystem; assign different groups the responsibility for finding out more about different system components. These components should include:

- plants commonly found in small freshwater streams in your area
- animals commonly found in and around freshwater streams in your area
- microscopic life common to small freshwater streams in your area

5. Ask them how they plan to find out about these elements. If possible, suggest a visit to a local stream for direct observations.

THINGS TO ASK

1. *Problem-Related Questions*:
   - How does this new information change the problem? How will this situation change the problem?
   - Are the ecology group's concerns legitimate?
   - What information do we already know that will help us answer the ecology group's questions?
   - What could be happening to the plants around the stream? To the soil? To the creatures in the stream?
   - Would the effects be long-term or short-term?
   - How will you find out what could be happening?
   - What could be found out through research? Who could we call?
   - Are there experiments we could conduct to find out?
   - How will you respond to the ecology group?
   - What's the best way to organize the group to find the answers to these new questions?

2. *Ecosystem-Related Questions*:
   - What should the boundaries of the creek ecosystem be? Why?
   - What are the abiotic elements present in the creek ecosystem? What are the biotic elements of the ecosystem?
   - Should we call the acid an element of the ecosystem or input into the ecosystem? Why?

- What are some interactions between biotic elements in the ecosystem? (Possible suggestions: predator-prey relationships, food chains in general.)
- Is the water pH a biotic element or an abiotic element?
- How can we find out more about stream ecosystems?

Handout 8.1

System Parts Chart

1. What are the boundaries of the system? Why did you choose them? Were there other possibilities?

2. List some important elements of the system.

3. Describe input into the system. Where does it come from?

4. Describe output from the system. What part(s) of the system produce it?

5. Describe some important interactions:

   - among system elements

   - between system elements and input into the system

6. What would happen to the system if the interactions in 5a could not take place? In 5b?

HANDOUT 8.2
SYSTEM DIAGRAM

Draw a picture of the system. Show where its boundaries are. Label and describe all of its parts.

HANDOUT 8.3

COMMUNIQUE FROM THE FRIENDS OF (_____*)

To: Highway Patrol Supervisor
From: Central Office

Some econut group calling itself "The Friends of _____*" has sent us the enclosed FAX. They claim that if we don't get this mess cleaned up ASAP, they'll alert the national media and start a boycott of Coleman Chemicals. They are also saying that they are preparing to sue the state for improperly handling the spill and sue Coleman Chemicals for transporting HCl on a public highway. How's the cleanup coming???

ALERT! ALERT! ALERT! ALERT!

_____* is getting more and more polluted by the minute . . . its water is getting more and more dangerous for people and other living things—and nobody is doing anything! Call your elected representatives! Call the highway patrol! Let them know how you feel about dangerous materials being transported on your highways! Put the pressure on and SAVE _____*!

Signed,

The Friends of _____*

---

* Insert the name of your water source—Example: Communique from The Friends of Clear Creek.

Session 2

THINGS TO DO

1. Have each group report on its findings.
2. Ask the different groups to think about what would happen to their organisms if acid spilled into the creek. Have each group make a list dividing their organisms into two groups: those that would probably be seriously affected by an acid spill and those that would probably not suffer from an acid spill. Have them report their predictions to the class and justify them.
3. Based on the findings of the small groups, have students fill out a new copy of the Systems Parts Chart (Handout 8.1). Also have them diagram their findings by drawing a System Diagram (Handout 8.2).
4. Ask students to answer the questions from the Problem Log (Handout 8.4) in their Problem Logs.

THINGS TO ASK

- Where do your organisms live?
- What do your organisms eat?
- What do your organisms drink?
- Do your organisms need anything special to survive?
- What interactions with other organisms does your organism need to survive?
- What output does your organism produce?
- What input into the system does your organism need?
- How could you find out if your organisms would be adversely affected by acid?
- Could you do an experiment to find out?
- Would it be ethical to do such an experiment? (Probably yes for plants and microscopic organisms, probably no for large mammals.)

ASSESSMENT

All worksheets may be used to assess student progress.

EXTENSION

Organize a field trip to a nearby creek or small pond to observe the freshwater aquatic ecosystem directly. Take along collecting equipment for insects, plants, and microscopic life; have students record as many observations about larger animals that are present as possible. Using field guides, identify as many of the organisms present as you can. Discuss your results in class.

NOTE TO TEACHER

If you have localized the problem, reproduce the Communique (Handout 8.3) so that the "Friends" represent the body of water affected by your fictional acid spill.

HANDOUT 8.4
PROBLEM LOG QUESTIONS

1. Describe your favorite organism in the set of organisms you researched. How does it make its living?

2. How could you find out whether your organism would be harmed directly by the addition of acid to the creek? Would it be ethical to do an experiment to find out? Why or why not?

3. Do you think that your favorite organism would be harmed indirectly by the addition of acid to the creek? Why or why not?

Lesson 9

Measurement of Water Flow

LESSON LENGTH: Two sessions

INSTRUCTIONAL PURPOSE

- To provide students the opportunity to measure and compute the rate of the spill and the water flow in the creek and decide its relevance to the problem.

MATERIALS AND HANDOUTS

Bucket
Graduated cylinder (10 ml)
Graduated cylinder (100 ml)
Pyrex liquid measuring cups (1 pint, 1/2 pint)
Stopwatch
Funnel
Optional rubber tubing that can be attached to the funnel (6 feet)

Handout 9.1: Student Brainstorming Worksheet
Handout 9.2: Student Experiment Worksheet
Handout 9.3: Student Protocol Worksheet
Handout 9.4: Laboratory Report Form
Handout 9.5: Problem Log Questions

THINGS TO DO

1. Ask students whether it might be useful to know how fast the acid is dripping into the creek and how fast the creek is flowing. Why would this information be important? How could they measure flow rates? Could this be modeled in the lab?

2. Set up a slow drip from a faucet. Show it to students. Tell them that they need to come up with a way to measure how much water flows from the faucet per hour when it is dripping.

3. Show them the materials for this lesson. Break students into small groups, and ask them to come up with a method for measuring the water flow rate from the faucet using the materials available. Tell them that they need to be able to write this method down in such a way that other people could repeat it. Pass out a copy of the Student Brainstorming Worksheet (Handout 9.1) and have them fill it out. When they have finished brainstorming, discuss their ideas.

4. Pass out copies of the Student Experiment Worksheet (Handout 9.2) and have students fill it out. Discuss their finished worksheet with them.

5. After students have filled out the Student Experiment Worksheet (Handout 9.2), pass out the Student Protocol Worksheet (Handout 9.3) and have each group write their experimental protocol. If time is running short, this could be assigned as homework.

NOTE TO TEACHER

Sample Protocol for Measuring Water Flow

1. Attach the rubber tubing to the bottom of the funnel; place the other end of the rubber tubing in the bucket.

2. Set the stopwatch for 5 minutes; put the funnel under the drip and collect water for exactly five minutes (OR place a measuring cup under the faucet instead of tubing and funnel).

3. Make sure that all of the collected water is in the bucket, not left in the tubing or the funnel. Measure the water in the bucket using either the graduated cylinders or the Pyrex measuring cups, depending on how much or how little water was collected.

4. Repeat the experiment several times, recording the volume of water collected each time.

5. Average the results; multiply by 12 to get the amount of water dripping from the faucet per hour.

Handout 9.1
Student Brainstorming Worksheet

1. What do we need to find out? (What is the scientific problem?)

2. What materials do we have available?

3. How can we use these materials to help us find out?

4. What do we think will happen? (What is our hypothesis?)

5. What will we need to observe or measure in order to find out the answer to our scientific question?

Adapted from Cothron, J. G., Giese, R. N., & Rezba, R. J. (1989). *Students and research.* Dubuque, IA: Kendall/Hunt Publishing Co.

HANDOUT 9.2
STUDENT EXPERIMENT WORKSHEET

Title of Experiment:

Hypothesis (Educated guess about what will happen):

Independent Variable (The variable that **you change**):

Dependent Variable (The variable that responds to changes in the independent variable):

Observations/Measurements to Make:

Constants (All the things or factors that remain the same):

Control (The standard for comparing experimental effects):

Handout 9.3

STUDENT PROTOCOL WORKSHEET

1. List the materials you will need.

2. Write a step-by-step description of what you will do (like a recipe!). List every action you will take during the experiment.

3. What data will you be collecting?

4. Design a data table to collect and analyze your information.

Session 2

THINGS TO DO

1. Check the protocols for safety and workability and have students perform their experiments.
2. Discuss their results.
3. Have students fill out a Laboratory Report (Handout 9.4) and incorporate it into their Problem Logs.
4. If this experiment was performed by a small group rather than by the whole class, have the group members report their findings to the class as a whole. Have the small group provide copies of all of their worksheets and their data to the other students. Have all students answer the questions in Handout 9.5 in their Problem Logs.

THINGS TO ASK

1. *Experiment-Related Questions*:
   - Did every group get the same answer? Why or why not?
   - Did anything go wrong?
   - How would you change your protocol next time to improve its effectiveness?
   - If you had access to other kinds of equipment, could you do a better job of measuring water flow? What additional equipment would help, and why?

2. *Problem-Related Questions*:
   - What might this have to do with the acid spill?
   - What liquids are flowing at the acid spill site?
   - Why might we want to measure the acid flow rate? The flow rate of the creek?
   - Could we use the methods we worked out in this experiment to measure the flow rate of the acid? Of the creek?
   - Would the methods we worked out be safe to use around an acid spill? How might we have to change them?
   - How do people really measure the rate of flow of creeks or rivers?
   - Does this make us need to modify the "Need to Know" board?

Assessment

1. Problem Log Questions.
2. Experimentation Handouts.

Extension

Students could measure the flow rate of a real stream, if one is available. As a math extension, students could do the following experiment:

Using a real stream or a simulated stream, students could release ping-pong balls into the stream and determine the amount of time needed for the stream to transport them a set distance. Students could then measure the width and depth of the channel. The flow rate will then be $\frac{distance}{time}$ x width x length.

Students could draw and label a cross-section of the stream.

HANDOUT 9.4

LABORATORY REPORT FORM

---

1. What did you do or test? (Include your experiment title)

2. How did you do it? What materials and methods did you use?

3. What did you find out? (Include a data summary and the explanation of its meaning)

4. What did you learn from your experiment?

5. What further questions do you now have?

6. Does the information you learned help with the problem?

HANDOUT 9.5

PROBLEM LOG QUESTIONS

1. Why would it be important to know the flow rates of the creek and of the dripping acid in the problem?

2. Given the flow rate that was calculated from the dripping faucet:
   a. How much water flows from the faucet every minute?

   b. How much water flows from the faucet every day?

   c. How much water would flow from the faucet in a year?

3. In a year, how many Olympic-size swimming pools could be filled by this dripping faucet?

Lesson 10: The Effect of Acid on Plants

LESSON LENGTH: One session to plan; one session for setup; three weeks of brief observation time; twenty minutes of discussion of results

INSTRUCTIONAL PURPOSE
- To explore the effects of acid on plants in order to help students determine possible implications of acid spill on the creek ecosystem.

MATERIALS AND HANDOUTS

Large cups (4 per group)
Many potted plants of same type and size (for example, flats of tomato seedlings or green onions)
White vinegar
Distilled water
Liquid measuring devices (measuring cups, graduated cylinders, pipettes)
Containers for liquids

Handout 10.1: Student Brainstorming Worksheet
Handout 10.2: Experiment Worksheet
Handout 10.3: Student Protocol Worksheet
Handout 10.4: Laboratory Report Form
Handout 10.5: Problem Log Questions

Session 1

THINGS TO DO

1. As students are responding to the "Communique from the Friends of (your water source)," one question that should come up is the question of what will happen to the organisms in and around the creek. Discuss this question with students.

115

2. Show students the materials listed above and ask them how they could use them to find out what effects an acid spill would have on living things.

3. Pass out a copy of the Student Brainstorming Worksheet (Handout 10.1) and have them fill it out; when they have finished brainstorming, discuss their ideas.

4. Pass out copies of the Student Experiment Worksheet (Handout 10.2) and have students fill it out. Discuss their finished worksheet with them.

5. After students have filled out the Student Experiment Worksheet (Handout 10.2), pass out the Student Protocol Worksheet (Handout 10.3) and have each group write their experimental protocol. If time is running short, this could be assigned as homework.

NOTE TO TEACHER

Sample Protocol For The Effect of Acid on Plants

1. The class will be broken into twelve lab groups of two people each. Each lab group will take green onions and four large cups from the table in the front of the room. Each group will trim the onions' roots to a length of 1 cm, then place a green onion in each cup. The cups will then be labeled as follows:

    #1: No acid (control)
    #2: white vinegar
    #3: 1:10 dilution of white vinegar
    #4: 1:100 dilution of white vinegar

2. Add 100 ml of the appropriate solution to each cup. Record observations about the color and appearance of each plant.

3. Place the cups in a place where they receive indirect natural light.

4. Every weekday for three weeks, measure and record the length of each plant's roots. Record any observations about the plant's color and appearance. Every Monday, Wednesday, and Friday, replace the liquid in each cup with 100 ml of the appropriate solution.

5. Pool all length data with those of the other lab groups; graph the data; discuss the recorded observations and the length results.

    A sample data table based on this protocol would look like the following:

Date	Plant	Length of Roots	Observations
1/7	control	12 cm	green, healthy
1/7	vinegar	13 cm	green, healthy

Handout 10.1
Student Brainstorming Worksheet

1. What do we need to find out? (What is the scientific problem?)

2. What materials do we have available?

3. How can we use these materials to help us find out?

4. What do we think will happen? (What is our hypothesis?)

5. What will we need to observe or measure in order to find out the answer to our scientific question?

Adapted from Cothron, J. G., Giese, R. N., & Rezba, R. J. (1989). *Students and research.* Dubuque, IA: Kendall/Hunt Publishing Co.

Handout 10.2
STUDENT EXPERIMENT WORKSHEET

Title of Experiment:

Hypothesis (Educated guess about what will happen):

Independent Variable (The variable that **you change**):

Dependent Variable (The variable that responds to changes in the independent variable):

Observations/Measurements to Make:

Constants (All the things or factors that remain the same):

Control (The standard for comparing experimental effects):

HANDOUT 10.3
STUDENT PROTOCOL WORKSHEET

1. List the materials you will need.

2. Write a step-by-step description of what you will do (like a recipe!). List every action you will take during the experiment.

3. What data will you be collecting?

4. Design a data table to collect and analyze your information.

Session 2

THINGS TO DO

1. Check the protocols for safety and workability and have students set up their experiments. Over the next three weeks, allow them a few minutes every session to make observations.

2. Once the experiments have run their course, discuss the results.

3. Have students fill out a Laboratory Report Form (Handout 10.4) and incorporate it into their Problem Logs.

4. If this experiment was performed by a small group rather than by the whole class, have the group members report their findings to the class as a whole. Have the small group provide copies of all of their worksheets and their data to the other students. Have all students answer the questions in Handout 10.5 in their Problem Logs.

THINGS TO ASK

1. *Systems questions*:
   - What system(s) are you studying in this experiment?
   - What are their boundaries and elements?
   - During the experiment, was there any input into the system?
   - Was there any output?
   - How did the system(s) change with time?
   - How do your results relate to the effects of acid input into the stream ecosystem defined in the problem?

2. *Data analysis questions*:
   - Did any changes occur in any of the dependent variables?
   - Were the changes the same in the experimental plants and the control plants, or were they different?
   - Are these changes important?
   - Did everyone get the same kinds of results, or were there differences? What does that tell you?

3. *The scientific question*:
   - Based on your data, what are the effects of acid on plant growth?
4. *Problem-related questions*:
   - Based on your results, what do you think the acid spill would do to the plants along the creek?
   - How does this information affect a possible solution to our acid spill problem?
   - Does this bring up any more questions for our "Need to Know" board?

ASSESSMENT

1. Problem log entries related to experimental design and data collection
2. Answers to in-class and/or Problem Log Questions.

EXTENSIONS

1. This experiment can also be repeated using neutralized acid when students are looking at problem solutions to see what happens if the highway acid is neutralized first and then rinsed. A neutralized acid can be made with a mixture of white vinegar and baking soda; check the pH to be sure it's really neutral.
2. Perform similar experiments using pond water as the experimental substance (to look at the effect of acid on aquatic microorganisms).

NOTE TO TEACHER

You will need to plan this experiment two weeks in advance since the results of this experiment will take a while to obtain.

Handout 10.4

Laboratory Report Form

1. What did you do or test? (Include your experiment title.)

2. How did you do it? What materials and methods did you use?

3. What did you find out? (Include a data summary and the explanation of its meaning.)

4. What did you learn from your experiment?

5. What further questions do you now have?

6. Does the information you learned help with the problem?

HANDOUT 10.5
PROBLEM LOG QUESTIONS

1. What effect did the acid have on plants?

2. What do you think this means for the problem?

3. How would you test the effects of acid on other kinds of organisms? Choose an organism and describe a way of finding out what acid would do to it.

4. Would it be ethical to do this experiment? Why or why not?

Lesson 11: The Effect of Acid on Materials

LESSON LENGTH: One session for setup; several weeks for observations; half a session for discussion of results

INSTRUCTIONAL PURPOSE
- To allow students to explore the effect of acid on various materials.

MATERIALS AND HANDOUTS

Baby food jars or other clear, sealable containers

Various materials to be tested: glass, metal, fabric, asphalt, plastic, leaves, grass, chicken bones, eggs, concrete

White vinegar

Protective Covering: aprons, safety goggles, gloves

Session 1

THINGS TO DO

1. Look at "Need to Know" board and speculate on the effect of acid on various materials.

2. Show students the available materials. Have each student choose one item and write down as many observations about its properties as possible in their Problem Logs.

3. Then, have students immerse a sample of their material in white vinegar in a baby food jar. They should then tightly cap the jar and write down observations about the appearance of the material immediately after it is placed in the acid. They should also immerse a sample of their material in water in a baby food jar and tighten the cap; observations about the appearance of the material in the water should also be recorded.

4. Each day, students should observe the jars and write down any changes in the appearance of their materials in their Problem Logs.

5. After two weeks, students should open the jars and discard the liquid. They should then rinse the materials under tap water and compare them to untreated samples of the same materials. Finally, they should observe and record the differences between the acid-treated and the untreated materials.

6. Each student should report on his/her results; class discussion should focus on the differences seen with different materials.

THINGS TO ASK

1. *Before the experiment:*
   - Why is it important to know the effect of acid on different materials?
   - What materials would be important to know about for the problem?
   - How could we use these materials to test the effects of acid on materials?
   - How long should we observe the materials in the acid? Do you think that changes will be instantaneous, or will they take time?

2. *After the experiment:*
   - What happened to the materials?
   - Did they all change the same way?
   - What might this mean?
   - Why did we put samples in water as well as vinegar?
   - Does this information change the way you think about the problem? If so, how? If not, why not?
   - What new questions does this raise in your mind?
   - We used white vinegar rather than HCl for safety reasons. How might the effects of HCl differ from those of vinegar? How could we find out?

ASSESSMENT

Explanation/description of changes in materials over time.

Lesson 12: Consultation with a Hazardous Waste Expert

LESSON LENGTH: Three sessions

INSTRUCTIONAL PURPOSE
- To provide interaction with a hazardous waste expert.

MATERIALS AND HANDOUTS
Chart and Markers
Audio-Visual equipment for guest speaker
Handout 12.1: Visitor Planning Sheet

Session 1: Before the Speaker Comes

THINGS TO DO

1. Brainstorm with students, deciding what questions need to be asked of the speaker. Use the "Need to Know" board to choose questions.

2. Class discussion can help sort questions into most and least important questions.

3. Students should also be guided to think about the best way to phrase the questions. Are they specific enough? Are they too specific?

4. Group questions can be recorded on a master question chart.

5. Students can then add any of their own questions to individual Visitor Planning Sheets (Handout 12.1).

THINGS TO ASK

- What information do we want to know?
- What information will the guest speaker be most qualified to give?
- What do we want to know by the time the guest speaker leaves?
- What facts do we want to get from this person?
- What opinions would be interesting to have?
- Which of these questions are most important?
- How can we get an idea of this person's perspective on this kind of situation?
- Do you think this person will have a bias? What would it be? How can we find out?

Session 2: The Guest Speaker's Presentation

THINGS TO DO

1. Guest Speaker: The guest provides his/her information regarding the area of his/her expertise.
2. Students take notes and ask their questions.
3. Students should also be prepared to share with the guest speaker background on the problem and their decisions to date.

Session 3: Debriefing

THINGS TO DO

1. In a follow-up to the guest speaker, teacher and students should review the "Need to Know" board, removing questions which have been answered and adding new issues, if necessary.
2. Teachers and students should discuss the potential bias in the information provided by the guest speaker and the possible effects of that bias on the validity of the information.

THINGS TO ASK

- What were the things we learned from the guest speaker?
- How does the new information affect our thinking about the problem?
- Do we need to reorganize our approach to the problem?
- Did this person reveal a particular bias? If so, what?
- Where can we go to get another perspective? A balanced report of information?

ASSESSMENT

1. Students should report in their problem logs information provided by the guest lecturer and reflect on the potential of bias in the problem log.
2. Students write a thank-you letter to the guest speaker, detailing which information was particularly helpful.

NOTE TO TEACHER

If the expert comes to the classroom, all students can participate. This format can also be used by small groups who need to interview an outside expert outside of class; afterwards, they can report any new information to the class.

HANDOUT 12.1
VISITOR PLANNING SHEET

Student Name _____

Name of visitor? _____

Who is this visitor?
_____
_____
_____

Why is this visitor coming to see us?
_____
_____

Why is this visitor important to us?
_____
_____
_____

What would you like to tell our visitor about our problem?
_____
_____

What questions do you want to ask the visitor?
_____
_____
_____

Lesson 13: Transport Regulations

LESSON LENGTH: Two–four sessions

INSTRUCTIONAL PURPOSE
- To facilitate the investigation of regulations governing the handling and transport of hazardous materials.

MATERIALS AND HANDOUTS
Handout 13.1: Problem Log Questions

THINGS TO DO

1. Use the "Need to Know" board to organize the information that students will have to research concerning the transportation of hazardous materials.

2. Divide questions among groups of students and have them brainstorm ways to find out the answers to their "Need to Know" questions. Their list ought to include library research, calling state transportation agency, calling a local trucking company, asking class resource people, and asking a lawyer.

3. Give students one or two sessions to research answers to their questions and work in small groups to prepare their findings. Alternatively, assign this work as homework. After research is complete, reconvene as a large group to share answers and synthesize the information to create a "big picture."

THINGS TO ASK

- Is it legal to transport hydrochloric acid?
- Does it have to be in a certain type of truck?
- Are there any regulations governing transportation?

- Where can we explore transportation regulations?
- Do the Friends of (your water source) have a right to sue?
- Do you think they should sue?
- Do you think they'll win the case? Why or why not?
- Compare the information that came from different sources. Was the information consistent?
- How did opinions about the regulations differ?
- Who do you think is right? Which position is most important?
- Would it be possible to change the regulations to serve the needs of all parties?

ASSESSMENT

Answers to questions in Handout 13.1.

ALTERNATIVES AND EXTENSIONS

Teachers interested in having students pursue the social science connections in more depth could have students try to revise the regulations to make them "better."

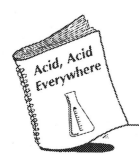

Handout 13.1
Problem Log Questions

1. What did you learn about transport regulations for hydrochloric acid?

2. How does this information affect the Friends of (your water source) lawsuit?

3. How does this information affect our problem?

Lesson 14: Resolution and Discussion

LESSON LENGTH: Five–six sessions

INSTRUCTIONAL PURPOSE
- To facilitate the development and selection of a solution to the acid spill problem.
- To provide a forum where students can have their solution reviewed by an audience of professionals in the field.

MATERIALS AND HANDOUTS
Handout 14.1: Weather Report
Handout 14.2: Problem Logs

Session 1

THINGS TO DO

1. Begin discussing possible problem resolutions with students; start by asking them to identify what the desirable solution would look like. Make sure that students have considered the question of "desirable" solution from many different perspectives, including those of commuters, Coleman Chemicals, and The Friends of (Clear Creek).

2. Discuss the question of values or priorities in achieving a solution. Which would be most effective in terms of cost and time?

3. Divide students into three groups to begin working out alternative solutions.

THINGS TO ASK

- What will a successful solution look like?
- Who will have to be satisfied?
- Who are the parties involved in the problem?
- What are the desired outcomes for the different parties?
- Do they all want exactly the same outcome? What are similarities? Differences?
- What are the possible short- and long-term issues?
- What do the optimal approaches look like?

Session 2

THINGS TO DO

1. Present students with the Weather Report (Handout 14.1).
2. Ask students how the Weather Report affects their approach to finding a resolution to the problem. Discuss any changes they may want to make in their strategy to get a solution.

THINGS TO ASK

- How does this change things?
- Does this solve the problem? Does it create another problem?
- Will four more days with acid in the area make a difference?
- Is dilution the best solution? What will the secondary effects be?
- What will the effect of the increased wind be?
- What new solution alternatives does this present?
- Which options are no longer possible?

Sessions 3–4

THINGS TO DO

1. Allow students to continue their work on their resolution, providing whatever resources they cannot get on their own. Guest speakers might be useful at this time, including chemists, ecologists, and decision strategists.

2. As teachers work with students in small groups they should remember to encourage students to think about and justify the relative weight they are giving to some issues over others.

Session 5

THINGS TO DO

1. Have each of the small groups present their solution to the problem. Each student group should include the criteria they selected to evaluate the success of the solution and their justification for selecting that criteria. Students should also rank order the relative importance of each criterion and discuss why some criteria ended up being more important than others.

2. Students should debate the merit of the proposals and either select one (with minor revisions allowed), combine elements of the same solutions, or devise a new solution to be presented to a professional audience for review. If necessary, a proposal can be sent with a minority report attached describing the viewpoint of a vocal minority.

Session 6

If possible, a panel of interested professionals representing various parties involved in the problem should be invited to class to hear students' solution. The panel may include community members who were consulted during the unit such as a hazardous waste expert, an ecologist, or lawyer. Otherwise, the solution should be mailed to one of them with a polite request for feedback.

ASSESSMENT

1. Group presentations.
2. Answers to Problem Log (Handout 14.2).

NOTE TO TEACHERS

Coach your panelists to ask questions of the students so that they must defend their position.

HANDOUT 14.1

FROM THE CHANNEL 7 WEATHER CENTER

THE LONG RANGE FORECAST:

Tuesday—high temperature in the 60's, low in the 40's; clear skies, winds from the northwest at 10 miles an hour.

Wednesday—starting out the same as Tuesday, with highs in the mid-60's, clouding over in the afternoon and winds shifting to the southwest. 20% chance of showers. Low temperatures in the mid-40's.

Thursday—cloud cover increasing as the day progresses. Highs in the upper 60's, with winds from the southwest at 20-30 miles an hour. Chance of heavy thunderstorms in the afternoon. Chance of rain 30% in the morning; increasing to 80% in the afternoon.

Friday—heavy thunderstorms likely, with strong winds from the southwest at 40 miles an hour. Highs in the low 70's; chance of rain 90%.

HANDOUT 14.2
PROBLEM LOG QUESTIONS

1. What is your solution? Why? How did you come to this conclusion? Is it more important that the solution be fast, thorough, or inexpensive? Why?

2. What are the criteria that you will use to judge the success of your solution? Briefly describe why each criterion represents an important concern in the development of the solution.

   a.

   b.

   c.

   d.

   e.

   f.

   g.

   h.

   i.

   j.

3. Look at your criteria and look at the solution your group came up with. Did all of the criteria receive equal weight in the solution? Did some criteria seem to work against each other? What sacrifices did you have to make in order to come up with the "best possible solution?" Was your "best possible" solution different from your idea of an ideal solution? How?

Lesson 15: Final Overall Unit Assessment Activity

INSTRUCTIONAL PURPOSE
- To assess understanding of the scientific content taught by this unit.
- To assess the ability of the student to use appropriate scientific process skills in the resolution of a real-world problem.
- To assess student understanding of the concept of systems.

ESTIMATED TIME
The content assessment should take the students approximately thirty minutes; the experimental design assessment should take the students approximately thirty minutes; and the systems assessment should take the students approximately thirty minutes.

TAILORING THE ASSESSMENTS TO REFLECT STUDENT EXPERIENCES
The list of substances in Part C of the content assessment may differ from the list of substances that your students tested in lab. Be sure that it is changed to include ten substances that your students tested.

MATERIALS AND HANDOUTS

Handout 15.1: Final Content Assessment
Handout 15.2: Experimental Design Assessment
Handout 15.3: Systems Assessment
Scoring Protocols for Final Content Assessment, Experimental Design Assessment, and Systems Assessment

PROCEDURE
Have students complete assessments found in Handouts 15.1, 15.2, and 15.3.

Handout 15.1

Final Content Assessment (30 minutes)

1A. Draw the pH scale and label its parts.

1B. On your pH scale drawing, show where the following substances belong: highly acidic substances, neutral substances, and highly basic substances.

1C. These are some substances whose pH values you have tested in the lab: white vinegar, apple juice, orange juice, baking soda stock solution, Pepsi, shampoo, distilled water, tap water, creek water, and lemon juice. Under your pH scale drawing, write the names of each substance. Draw lines from each name to the right place on your pH scale drawing to show the pH of each one.

2. Suppose your teacher gave you a jar filled with a clear liquid and asked you to find its pH. Describe the materials you would need and the method you would use.
Materials:

Method:

3. How would you change the pH of your clear liquid from acidic to neutral? List the two different ways to do this.

   a.

   b.

4. What safety equipment and precautions would you need to handle your clear liquid safely?

   Equipment:

   Precautions:

HANDOUT 15.2

EXPERIMENTAL DESIGN ASSESSMENT (30 MINUTES)

You are a high school student, proud to have the care of your neighborhood baby wading pool as your first real summer job. Soon, though, you run into problems. Your neighbors complain about the condition of the baby pool on weekends. After hot, sunny weekends, the pool is dirty and filled with algae. You know that the chemicals that you add to the pool to prevent the growth of algae are very basic and that pool pH is important, so you decide that maybe the problem on hot weekends is that the pH of the pool is getting too close to neutral. This might allow algae to grow.

1. What experiment could you do that would allow you to test this idea? In your answer, include the following:

   a. Your hypothesis:

   b. The materials you would need:

   c. The protocol you would use:

   d. A data table showing what data you would collect:

   e. A description of how you would use your data to decide whether your idea about the pool pH and the growth of algae was correct.

Handout 15.3

Systems Assessment (30 minutes)

You can think of the baby pool as a system.
1. List the parts of the system in the spaces provided below. Include boundaries, elements, input, and output.

   Boundaries:

   Elements:

   Input:

   Output:

2. Draw a diagram of the system that shows where each of the parts can be found.

3. On your diagram, draw lines (in a different color) showing three important interactions between different parts of the pool system. Why is each of these interactions important to the system? Explain your answer.

    a. Interaction #1:

    b. Interaction #2:

    c. Interaction #3:

Scoring Protocol
Final Content Assessment

1. **(20 points)**

   a. **(7 points)** Draw the pH scale and label its parts.

   b. **(3 points)** On your pH scale drawing, show where the following substances belong: highly acidic substances, neutral substances, and highly basic substances.

For parts 1a and 1b, students' drawings should resemble the drawing below. Give a half point for each numerical label that may be present. For part b, give one point each for correctly located labels.

The pH scale, labeled as required in parts 1a and 1b, is shown below.

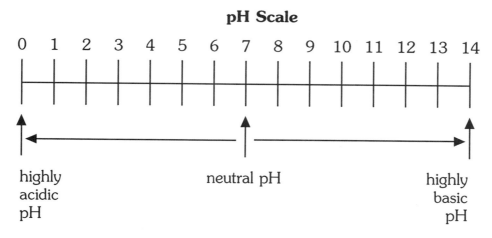

   c. **(10 points)** These are some substances whose pH values you have tested in the lab: white vinegar, apple juice, orange juice, baking soda stock solution, Pepsi, shampoo, distilled water, tap water, creek water, and lemon juice. Under your pH scale drawing, write the names of each substance. Draw lines from each name to the right place on your pH scale drawing to show the pH of each one.

Give one point for each correctly located substance, up to a total of ten points. Approximate locations for a variety of substances are given below.

(**Note:** The teacher grading this will need to make sure that the list of substances corresponds to substances actually tested by the students; the list should be altered appropriately if students tested different liquids than those listed. As the pH of some substances (such as tap water) varies depending on their source, give students credit for the answer that they obtained in lab *for these substances*.)

Acids: vinegar, lemon juice, apple juice, orange juice, hydrochloric acid, sulfuric acid, Pepsi

Neutral substances: distilled water (pH should be EXACTLY 7); students may also list water from other sources, such as pond water; some shampoos

Bases: baking soda stock solution, some shampoos

2. **(10 points total)** Suppose your teacher gave you a jar filled with a clear liquid and asked you to find its pH. How would you do this? Describe the materials you would you need and the method you would use.

    Materials: **(5 points)**

pH paper or a pH meter (litmus paper is not an acceptable answer, as it does not give a precise enough reading; neither is "an indicator," for the same reason); give 5 points for either correct answer

    Method: **(5 points)**

For either method, a correct answer will involve dipping the pH paper (or the pH meter probe) into the liquid, then reading the pH on the scale (done for pH paper by matching the color of the pH paper to the scale on the side of the box; done for a pH meter by reading the number off the meter's scale, or by reading the digital answer on the display). Give 5 points for any answer that combines the ideas of dipping the appropriate item into the liquid and then reading the answer off the appropriate scale or display.

3. **(10 points)** How could you change the pH of your clear liquid from acidic to neutral? List the two different ways to do this.

    a. **(5 points)** Dilution (adding water)

    b. **(5 points)** Neutralization (adding base)

In each case, give 5 points if the answer is correct; appropriate terminology is not necessary.

    Other answers, such as "letting it sit," are not acceptable.

4. **(10 points total)** What safety equipment and precautions would you need to handle your clear liquid safely?

    Equipment **(5 points)**: safety goggles or glasses, lab coat or apron, filter mask, gloves, fume hood, or any other reasonable piece of safety equipment.

    Precautions **(5 points)**: not tasting the liquid, not running in lab, not pipetting by mouth, no horseplay, careful handling procedures for the jar, or any other reasonable rule.

If even one correct piece of equipment is listed, give five equipment points; if even one reasonable safety rule is listed, give five precaution points.

Total number of points possible: 50

SCORING PROTOCOL
EXPERIMENTAL DESIGN ASSESSMENT

You are a high school student, proud to have the care of your neighborhood baby wading pool as your first real summer job. Soon, though, you run into problems. Your neighbors complain about the condition of the baby pool on weekends. After a hot, sunny weekend, the pool is dirty and filled with algae. You know that the chemicals that you add to the pool to prevent the growth of algae are very basic, and that pool pH is important, so you decide that maybe the problem on hot weekends is that the pH of the pool is getting too close to neutral. This might allow algae to grow.

1. What experiment could you do that would allow you to test this idea? In your answer, include the following:

    a. **(10 points)** Your hypothesis:
    If the pool pH becomes sufficiently close to neutral, algae can grow.

(**Note:** Other hypotheses are also acceptable, but they will require different experiments. This hypothesis is merely a sample.) *Give ten points for any reasonable hypothesis.*

   b. **(10 points)** The materials you would need:
   —*pH paper (and the scale on the side of the box)*
   —*the baby pool*

Give ten points for a simple materials list. This list does not need to be all-inclusive (notice that paper and writing implements are not included) but does need to contain any testing equipment required for the experiment.

(**Note:** If the students have come up with a different experiment than the one outlined here, accept their materials list if it looks reasonable.)

   c. **(10 points)** The protocol you would use:

On each day of each of the next six weekends, I would use pH paper to test the pH of the pool water in the baby pool and also see whether algae were present (or at least obvious to the naked eye, as slime in the pool). I would record these observations in the data table.

(**Note:** Accept any reasonable protocol, providing it is consistent with the student's ideas in parts a and b.)

Give ten points for a simple, general explanation of the protocol for the experiment. A detailed protocol, listing every step to be taken, is not required for this assessment.

d. **(10 points)** A data table showing what data you would collect:

    Date   Time   pool pH   Is slime present?   Y/N

Give ten points for a data table that includes headings for every measurement/observation planned by the student. The data table does not need to include sample data.

(Note: Accept any data table consistent with the student's planned experiment.)

e. **(10 points)** A description of how you would use your data to decide whether your idea about the pool pH and the growth of algae was correct.

Each time I collected information about the pool pH and the presence or absence of slime would be one data point. I would make a graph of these points showing presence of slime vs. pool pH. If it was clear from my graph that algae only appeared in the pool when the pH of the pool water fell below a certain level, I would say that my hypothesis had been confirmed. If there was no correlation between the pool pH and the presence of algae, I would say that my hypothesis was incorrect.

(Note: Accept any answer consistent with the student's planned experiment.)

Give five points for a clear description of how the data would be analyzed in order to come up with an answer; give five points for a clear description of how that answer would be used to determine whether the hypothesis was correct or incorrect.

Total number of points possible: 50

SCORING PROTOCOL
SYSTEMS ASSESSMENT

You can think of the baby pool in the Experimental Design Assessment as a system.

1. **(25 points)** List the parts of the system. Include boundaries, elements, input, and output.

Boundaries: **(10 points)**

The boundaries of this system are the walls and surface of the pool (including the drain but not the sewer).

(Note: *The kids may have different boundaries for this system. As long as their other answers in this section are consistent with these boundaries, accept their answers.)*

Give ten points for clearly specified boundaries. These boundaries should completely enclose the system; there should be no question about whether something is inside the system or not. If the boundaries described are only partially complete, give only five points; if no boundaries are described at all, give zero points.

Elements: **(5 points)**

Pool water, algal spores, bacteria, the surface of the pool walls, the drain and filter basket.

Give two points for a single reasonable item; give five points for any answer that has two or more reasonable items.

Input: **(5 points)**

Fresh tap water, rainwater, kids and their by-products, bacteria, viruses, suntan oil, dirt, rocks, pool toys, bugs, leaves, pool chemicals, sunlight, frogs . . .

Give two points for a single reasonable item; give five points for any answer that has two or more reasonable items.

Output: **(5 points)**

Pool water that splashes out of the pool, water that evaporates, oxygen made by the algae, dirty water that leaves the pool through the drains, water that leaves on the surfaces of kids . . .

Give two points for a single reasonable item; give five points for any answer that has two or more reasonable items.

2. **(10 points)** Draw a diagram of the system that shows where each of the parts can be found.

Give ten points for a drawing that includes all of the system components listed by the student in their answer to part 1; give five points for a drawing that includes only some of the system components listed in the student's answer to part 1.

3. **(15 points)** On your diagram, draw lines (in a different color) showing three important interactions between different parts of the pool system. Why is each of these interactions important to the system? Explain your answer.

   a. Interaction #1:

   Pool chemicals and algae: the pool chemicals keep the algae from growing and making the pool disgusting to use.

   b. Interaction #2:

   Kids and water: the kids splash the water out and lower the water level in the pool, which means that we have to add water to keep it from drying out.

   c. Interaction #3:

   Sunlight and algae: the algae need the sunlight to grow and eventually make the pool slimy.

   *(**Note:** Other reasonable examples of interactions should be allowed.) Give five points for each reasonable answer.*

Total number of points possible: 50

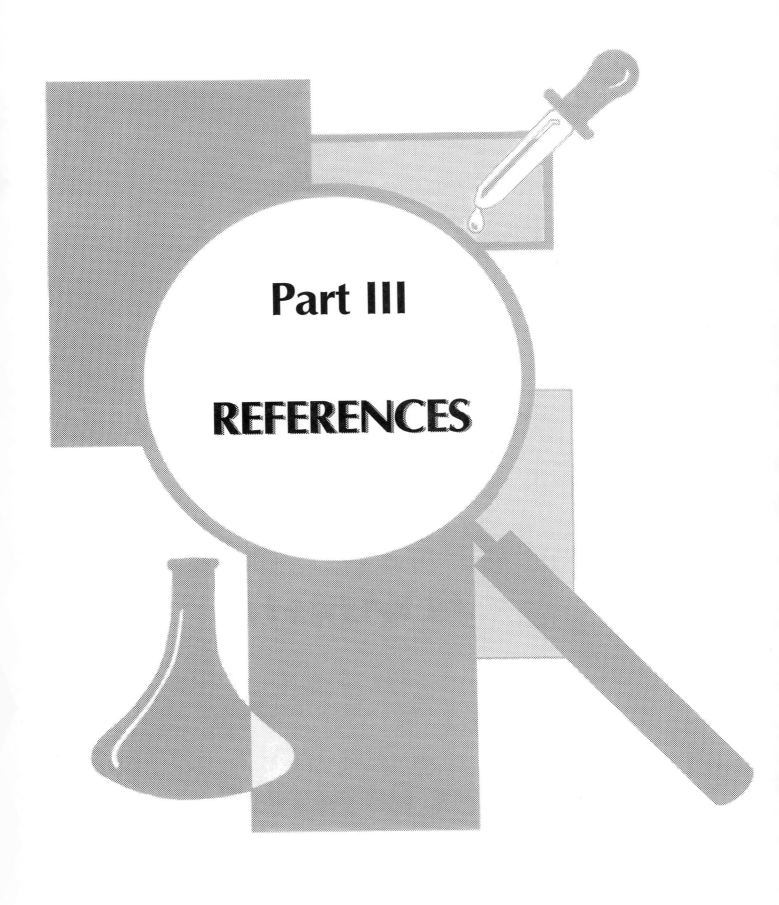

REFERENCES

Budavari, S. (Ed.) (1989). *The Merck Index: An encyclopedia of chemicals, drugs, and biologicals.* Rahway, NJ: Merck & Company.

Cothron, J.H., Giese, R.N., & Rezba, R.J. (1996). *Science experiments and projects for students.* Dubuque, IA: Kendall/Hunt Publishing Company.

Cothron, J.H., Giese, R.N., & Rezba, R.J. (1996). *Science experiments by the hundreds.* Dubuque, IA: Kendall/Hunt Publishing Company.

Cothron, J.H., Giese, R.N., & Rezba, R.J. (1996). *Students and research: Practical strategies for science classrooms and competition.* Dubuque, IA: Kendall/Hunt Publishing Company.

Modular Units That May Be Integrated with *Acid, Acid Everywhere*
- Acid Rain (Grades 6–10)
- Cabbages and Chemistry (Grades 4–8)
    Available from:
    Great Explorations in Math & Science (GEMS)
    Lawrence Hall of Science
    University of California
    Berkeley, CA 94720-5200
    (510) 642-7771
- Chemical Tests
- Land and Water
    Developed by:
    Science and Technology for Children (STC)
    Distributed by:
    Carolina Biological Supply
    2700 York Road
    Burlington, NC 27215
    (800) 334-5551
- Mixtures and Solutions
    Developed by:
    Full Options Science Systems (FOSS)
    Distributed by:
    Britannica Educational Corporation
    310 S. Michigan Avenue, 6th Floor
    Chicago, IL 60604
    (800) 554-9862

- Stream Tables
- Water Flow
    Developed by:
    Elementary Science Study (ESS)
    Distributed by:
    Delta Education
    P.O. Box 915
    Hudson, NH 03051-0915
    (800) 258-1302

- The Waste Hierarchy
- Investigating Hazardous Materials
- Chemical Survey and Solutions and Pollution
- Household Chemicals
    Developed by:
    CEPUP, Lawrence Hall of Science
    Distributed by:
    Innovative Learning Publications (Addison-Wesley)
    Route 128
    Reading, MA 01867
    (800) 552-2259

Order these outstanding titles by the
CENTER FOR GIFTED EDUCATION

SCIENCE

QTY	TITLE	ISBN	PRICE	TOTAL
	Guide to Teaching a Problem-Based Science Curriculum	0-7872-3328-5	$32.95*	
	Acid, Acid Everywhere	0-7872-2468-5	$32.95*	
	The Chesapeake Bay	0-7872-2518-5	$32.95*	
	Dust Bowl	0-7872-2754-4	$32.95*	
	Electricity City	0-7872-2916-4	$32.95*	
	Hot Rods	0-7872-2813-3	$32.95*	
	No Quick Fix	0-7872-2846-X	$32.95*	
	What a Find!	0-7872-2608-4	$32.95*	

LANGUAGE ARTS

QTY	TITLE	ISBN	PRICE	TOTAL
	Guide to Teaching a Language Arts Curriculum for High-Ability Learners	0-7872-5349-9	$32.95*	
	Autobiographies Teaching Unit	0-7872-5338-3	$28.95*	
	Literature Packets	0-7872-5339-1	$37.00*	
	Journeys and Destinations Teaching Unit	0-7872-5167-4	$28.95*	
	Literature Packets	0-7872-5168-2	$37.00*	
	Literary Reflections Teaching Unit	0-7872-5288-3	$28.95*	
	Literature Packets	0-7872-5289-1	$37.00*	
	The 1940s: A Decade of Change Teaching Unit	0-7872-5344-8	$28.95*	
	Literature Packets	0-7872-5345-6	$37.00*	
	Persuasion Teaching Unit	0-7872-5341-3	$28.95*	
	Literature Packets	0-7872-5342-1	$37.00*	
	Threads of Change in 19th Century American Literature Teaching Unit	0-7872-5347-2	$28.95*	
	Literature Packets	0-7872-5348-0	$37.00*	

Method of payment:
❑ Check enclosed (payable to Kendall/Hunt Publishing Co.)
❑ Charge my credit card:
  ❑ VISA   ❑ Master Card   ❑ AmEx

Credit Card No. _____
Exp. Date _____
Signature _____
Name _____

AL, AZ, CA, CO, FL, GA, IA, IL, IN, KS, KY, LA, MA, MD, MI, MN, NC, NJ, NM, NY, OH, PA, TN, TX, VA, WA, & WI add sales tax.

Add shipping: order total $26-50 = $5; $51-75 = $6; $76-100 = $7; $101-150 - $8; $151 or more = $9

Price is subject to change without notice. **TOTAL**

Address _____
City/State/ZIP _____
Phone No. (   ) _____
E-mail _____

KENDALL/HUNT PUBLISHING COMPANY
4050 Westmark Drive   P.O. Box 1840   Dubuque, Iowa 52004-1840
A16/mkk Q2 2005 01

Call (800) 228-0810 • Fax (800) 772-9165
Visit us at www.kendallhunt.com

An overview of the outstanding titles available from the

CENTER FOR GIFTED EDUCATION

A PROBLEM-BASED LEARNING SYSTEM FROM THE CENTER FOR GIFTED EDUCATION FOR YOUR K-8 SCIENCE LEARNERS

The Center for Gifted Education has seven curriculum units containing different real-world situations that face today's society, plus a guide to using the curriculum. The units are geared towards different elementary levels, yet can be adapted for use in all levels of K-8.

The goal of each unit is to allow students to analyze several real-world problems, understand the concept of systems, and conduct scientific experiments. These units also allow students to explore various scientific topics and identify meaningful problems for investigation.

Through these units your students experience the work of real science in applying data-handling skills, analyzing information, evaluating results, and learning to communicate their understanding to others.

A LANGUAGE ARTS CURRICULUM FROM THE CENTER FOR GIFTED EDUCATION FOR YOUR GRADES 2-11

The Center for Gifted Education at the College of William and Mary has developed a series of language arts curriculum units for high-ability learners.

The goals of each unit are to develop students' skills in literature interpretation and analysis, persuasive writing, linguistic competency, and oral communication, as well as to strengthen students' reasoning skills and understanding of the concept of change.

The units engage students in exploring carefully selected, challenging works of literature from various times, cultures, and genres, and encourage students to reflect on the readings through writing and discussion.

The units also provide numerous opportunities for students to explore interdisciplinary connections to language arts and to conduct research around issues relevant to their own lives. A guide to using the curriculum is also available.